◀ A MATERIAL WORLD ▶

It's RUBBER

KAY DAVIES and WENDY OLDFIELD

Wayland

A MATERIAL WORLD

It's Cotton It's Plastic
It's Glass It's Rubber
It's Metal It's Wood

Editor: Joanna Housley
Designer: Loraine Hayes

First published in 1993 by
Wayland (Publishers) Ltd
61 Western Road, Hove
East Sussex BN3 1JD, England

British Library Cataloguing in Publication Data
Davies, Kay
It's Rubber. – (Material World Series)
I. Title II. Oldfield, Wendy III. Series
678

ISBN 0 7502 0855 4

Typeset by Kalligraphic Design Ltd, Horley, Surrey
Printed and bound in Belgium by Casterman S.A.

Words that appear in **bold** in the text are
explained in the glossary on page 22.

IT'S RUBBER

Rubber is a very useful material that comes from rubber trees. It is found as a sticky liquid, and is then processed to become strong and stretchy. A lot of rubber is used to make tyres for cars, aeroplanes and bicycles. It is waterproof, so we can use it to make rainwear, wet suits and dinghies. Rubber is used when we want to seal something so that air cannot get in, such as food jars. The stretchy elastic in many of our clothes is made from rubber. Look around you and use this book to help you find other things that are made from rubber.

Rubber comes from the white, sticky **sap** of the rubber tree. This sap is called **latex.**

It runs from slits made in the **bark** into collecting cups.

Foam rubber is made by blowing air into wet latex. When it dries it is soft and spongy. It makes a comfortable filling for furniture.

Can you build your own hideaway with foam cushions?

The strong rubber line stretches then bounces back. It stops the **bungy jumper** from hitting the ground.

How many uses can you think of for an elastic band?

Some clothes are easy to pull on and off. The elastic in waistbands and cuffs makes them fit snugly, even when we are exercising.

Long **waders** keep the fisherman warm and dry.

Thick rubber soles on shoes and boots protect our feet and stop them from slipping.

The huge tyres help the tractor move easily over bumpy or muddy ground. The driver has a smooth, safe ride.

Electric wires are covered in rubber to make them safe.

Coloured rubber helps the **electrician** to tell which wires to use.

Hot-water bottles keep our beds warm on cold nights. The water will stay hot for a long time because heat does not pass through rubber easily.

A big
rubber ball
filled with
air is very
bouncy.

It is fun
to jump
along on it.

The rubber in a balloon is so thin it stretches easily. Be careful not to burst it!

You can pretend you are somebody else in a rubber mask. Masks are fun to wear at parties and **carnivals**.

What animal does this mask look like?

Young children like to play with bright, soft, squeezy toys. These rubber toys are easy to wipe clean.

Rubber is used to make toys and balls for dogs because it can be chewed but will bounce back into shape.

Rubber seals on jars and bottles make a tight fit for the lids. Air cannot get inside, so the food will stay fresh for a long time.

We can do all sorts of jobs wearing rubber gloves.

They move with our fingers and protect our skin.

Erasers help us keep our work neat and tidy. The soft rubber removes our mistakes but does not harm the paper. Do you have pencils with erasers on the end?

Look at a
bicycle and
find all
the rubber
parts.
Can you
decide
why they
are there?

Are there
any rubber
parts you
cannot see?

19

The dancer is practising on a rubber mat.
It is much softer than the wooden floor.

Rubber mats protect our bodies if we are dancing or
jumping in a gym.

The old tyre makes a strong, safe swing. There are no sharp edges and the tough rubber will last for years.

GLOSSARY

Bark The outer layer of a tree.

Bungy jumper Someone who jumps from a height. The person is attached to a long stretch of rubber, so that he or she bounces back.

Carnivals Fairs or celebrations, normally held out-of-doors.

Electrician A person who makes and mends electrical equipment.

Erasers Pieces of rubber used for rubbing out something that has been written with a pencil.

Latex The milky sap of the rubber tree.

Sap The liquid that carries food round a plant.

Waders Tall waterproof boots.

BOOKS TO READ

Materials by Kay Davies and Wendy Oldfield (Wayland, 1991)

My Balloon by Kay Davies and Wendy Oldfield (A & C Black, 1989)

Some books in Collins' Primary Science series may also be useful.

TOPIC WEB

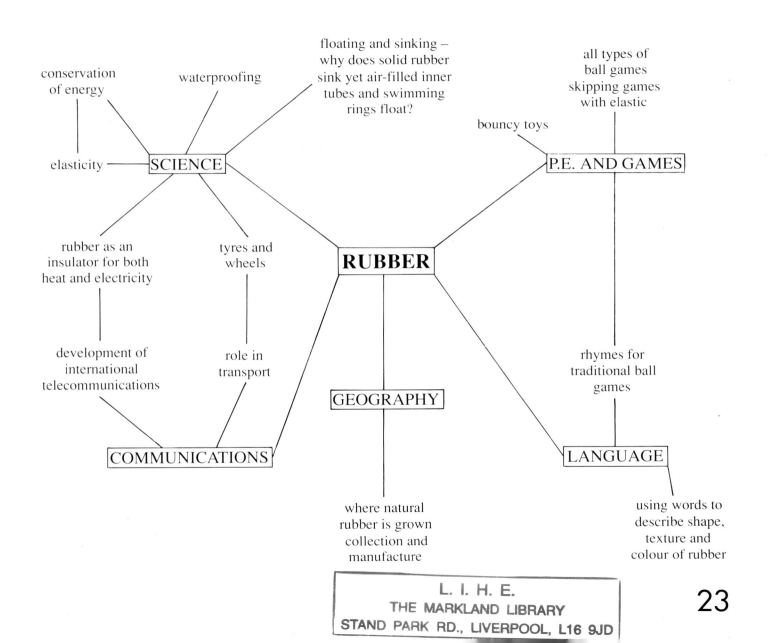

floating and sinking –
why does solid rubber
sink yet air-filled inner
tubes and swimming
rings float?

conservation
of energy

waterproofing

all types of
ball games
skipping games
with elastic

elasticity — SCIENCE

bouncy toys

P.E. AND GAMES

rubber as an
insulator for both
heat and electricity

tyres and
wheels

RUBBER

development of
international
telecommunications

role in
transport

rhymes for
traditional ball
games

COMMUNICATIONS

GEOGRAPHY

LANGUAGE

where natural
rubber is grown
collection and
manufacture

using words to
describe shape,
texture and
colour of rubber

INDEX

Picture acknowledgements

The publishers wish to thank the following for supplying the photographs in this book: Cephas Picture Library 16; Chapel Studios 6 (left), 8 (right), 11, 15 (left), 17; Bruce Coleman cover (left); Eye Ubiquitous cover (top), 6 (right, L Johnstone), 10 (D Cumming); Life File Photographic Library 5 (Pete Glastonbury), 12 (Nicola Sutton), 14 (Nicola Sutton), 15 (right, Tim Fisher); Tony Stone Worldwide 4 (Peter Gittoes); Wayland Picture Library 7, 8 (left), 13 (A Blackburn); ZEFA cover (right), 9, 18, 19, 20, 21.